OXFORD CONNECTIONS

LIGHT AND SHADOW

James Driver

Series editor Sue Palmer

OXFORD

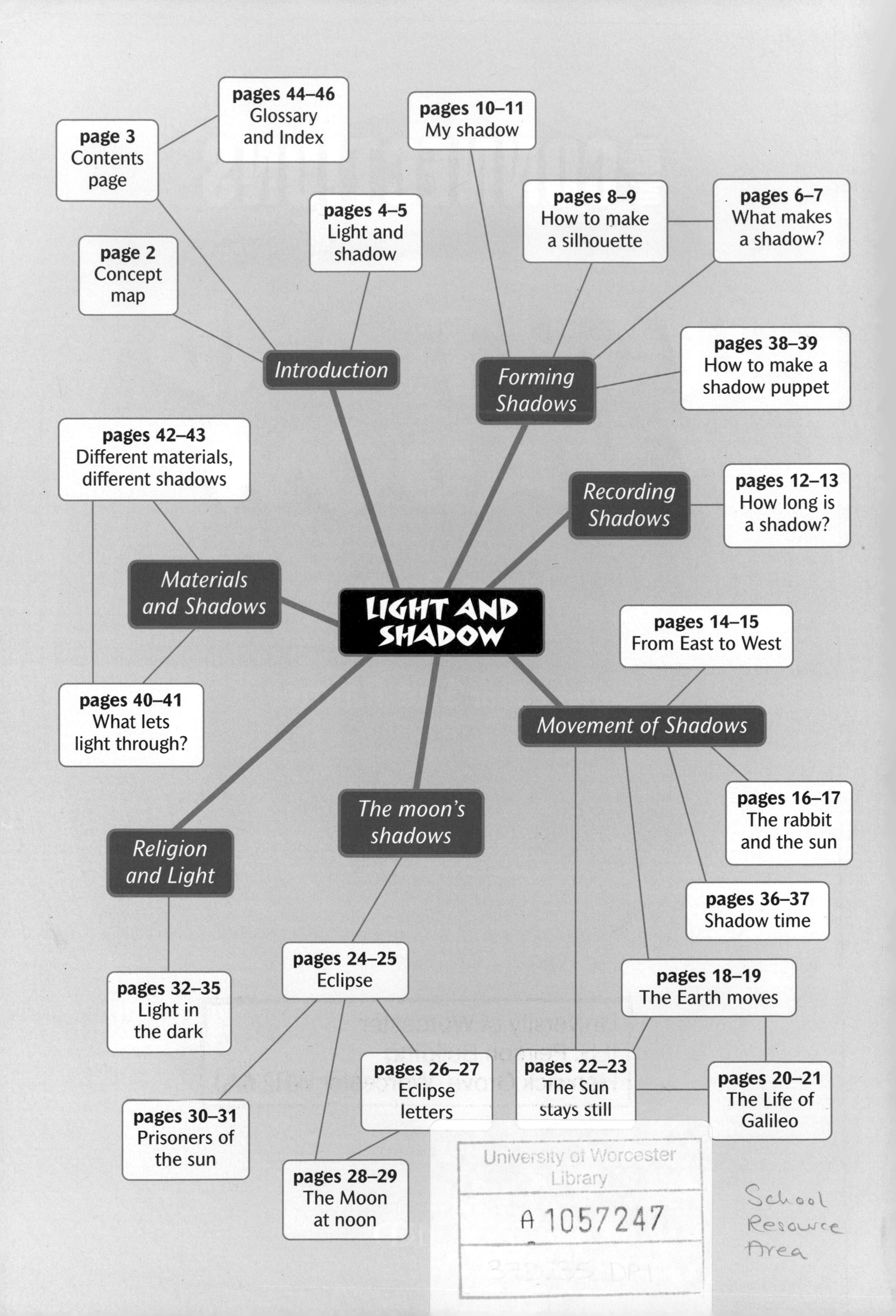
LIGHT AND SHADOW
Introduction
page 2 Concept map
page 3 Contents page
pages 4–5 Light and shadow
pages 44–46 Glossary and Index
Forming Shadows
pages 6–7 What makes a shadow?
pages 8–9 How to make a silhouette
pages 10–11 My shadow
pages 38–39 How to make a shadow puppet
Recording Shadows
pages 12–13 How long is a shadow?
Movement of Shadows
pages 14–15 From East to West
pages 16–17 The rabbit and the sun
pages 18–19 The Earth moves
pages 20–21 The Life of Galileo
pages 22–23 The Sun stays still
pages 36–37 Shadow time
The moon's shadows
pages 24–25 Eclipse
pages 26–27 Eclipse letters
pages 28–29 The Moon at noon
pages 30–31 Prisoners of the sun
Religion and Light
pages 32–35 Light in the dark
Materials and Shadows
pages 40–41 What lets light through?
pages 42–43 Different materials, different shadows

CONTENTS

The Sun is full of mystery, and people have always made up stories about it to try to explain what it does. You can read some of these stories in *'From East to West'* and in *'The Earth Moves'* you can also read some of the stories that scientists have told about the Sun.

Professor Kathy Sykes
Bristol University
Cheltenham Science
Festival Director

LIGHT AND SHADOW

We need light to see what is around us. It comes from many different **sources**. There are natural sources like the Sun, lightning and fire. There are also man-made sources of light, such as torches, lasers and television screens.

Light helps us in many ways. Natural light from the Sun allows us to see during the day. It also helps plants to grow and provides a source of heat. Man-made light helps us too. For instance, the powerful beams on lighthouses warn ships of dangerous rocks, and traffic lights make driving a car safer.

However, light can also be dangerous. Looking at the Sun can damage a person's eyes, and the Sun can also burn the skin if people stay outside in the sunshine for too long.

Light travels at 299,792 km per second!

Light travels very fast – at 299,792 km per second! However, it cannot always pass through objects that block its path.

Some materials, like brick, wood, silver foil and steel, block the light and stop it passing through. These are called **opaque** materials. When an opaque object blocks a beam of light, a shadow is formed. Other materials let the light pass through. These are **translucent** and **transparent** materials. Translucent materials include thin paper and fine cloth. Transparent materials include glass and clear plastic.

WHAT MAKES A SHADOW?

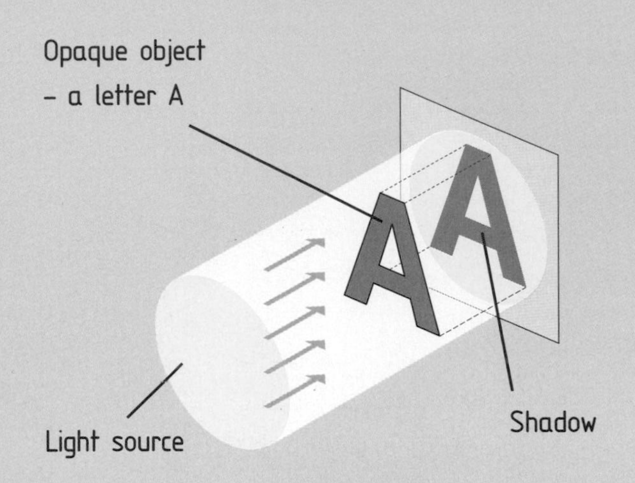

Light travels in straight lines

Beams of light travel in straight lines. This means that, unless someone interferes with a beam of light, it does not bend. Therefore it cannot go round an object that is blocking its path.

If the object is **opaque**, the light cannot go through it either. The area behind the opaque object stays dark, because the object blocks the light. This dark area is known as a shadow.

It does not make any difference where the light comes from. It might be moonlight, or light from the headlights of a car, a torch or a searchlight. Whenever light is blocked by an opaque object, it always produces a shadow.

The shadow always looks something like the shape that has blocked the light. However, if the light **source** is high above the object, the shadow is short; if the light source is close to the ground, the shadow is long.

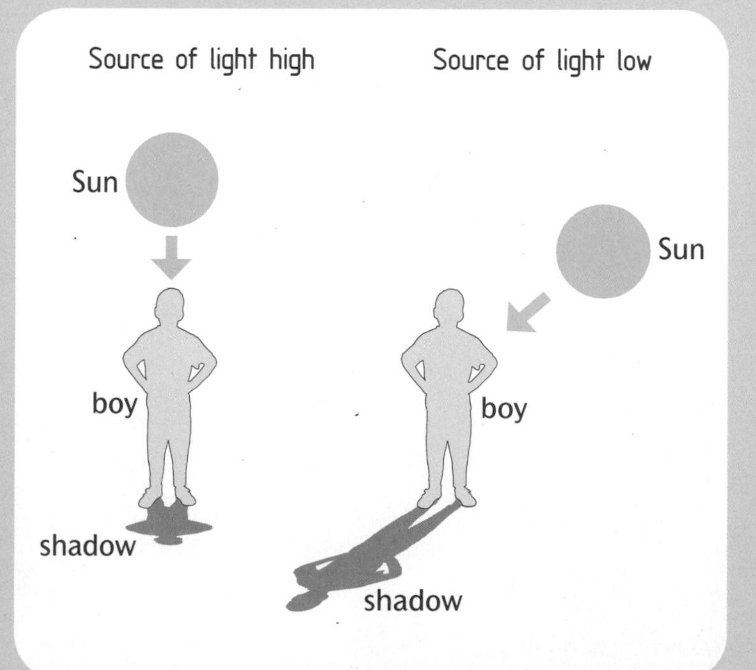

HOW TO MAKE A

Before cameras were invented, people used to make 'shadow pictures' of their friends and families. These shadow pictures are called **silhouettes.**

What you need

- A powerful light **source**, e.g. a torch or overhead projector
- A large sheet of white paper
- A large sheet of black paper
- Black marker pen
- White crayon or chalk
- Scissors
- Coloured paper for mounting the silhouette
- Glue

SILHOUETTE

1. Attach the white paper to a wall, so you can stand or sit in front of it.

2. Switch on the light source and point the beam towards the sheet of paper.

3. Stand sideways-on to the light source so that the shadow of your head is in the middle of the paper.

4. Ask a friend to draw around the shadow of your head with the black pen. Stand still!

5. Switch off the light source.

6. Cut around the outline and place it on the black paper.

7. Draw around the outline with white crayon or chalk. Cut it out carefully.

8. You have your silhouette. Glue it on to a piece of coloured paper to display.

MY SHADOW

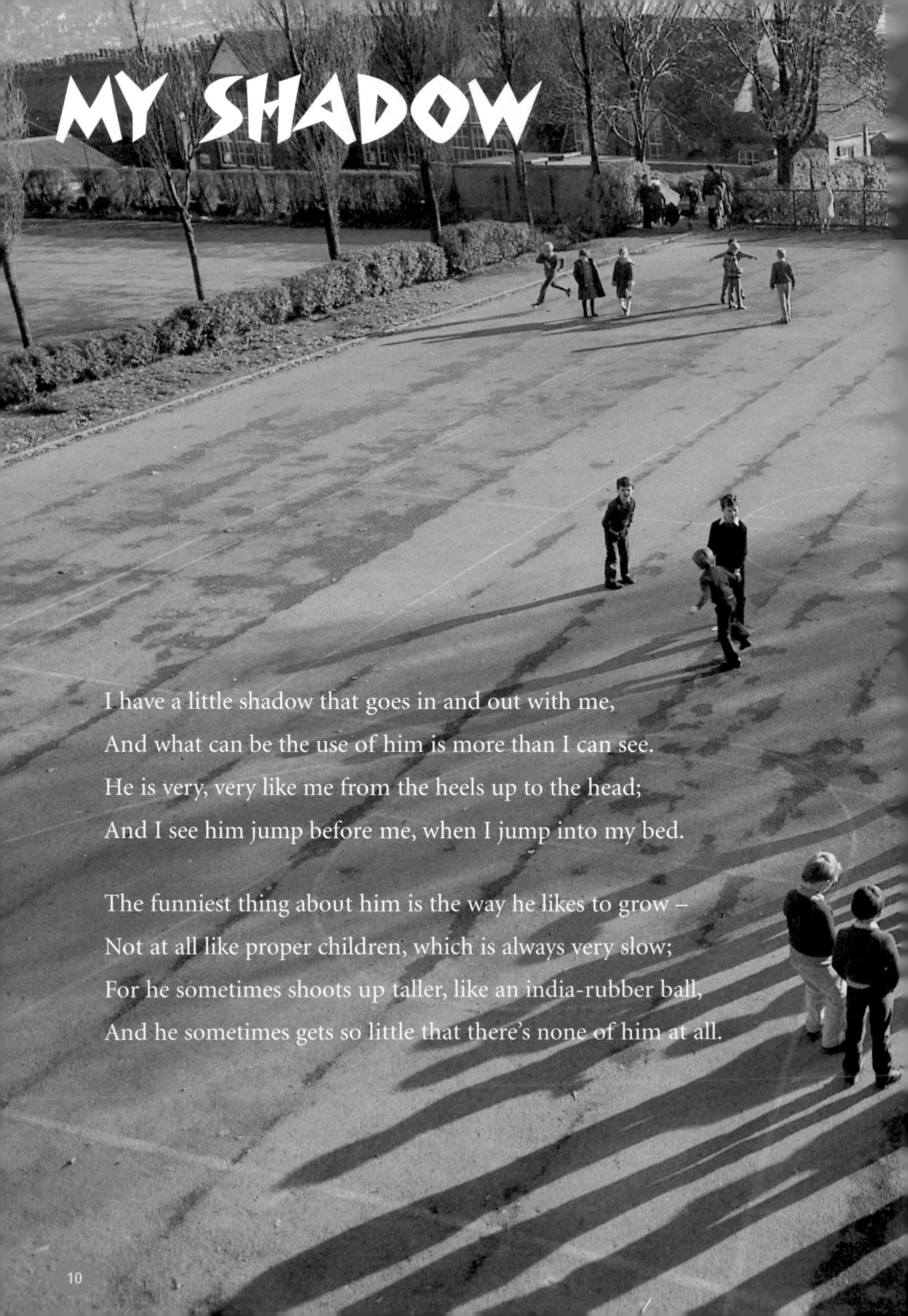

I have a little shadow that goes in and out with me,
And what can be the use of him is more than I can see.
He is very, very like me from the heels up to the head;
And I see him jump before me, when I jump into my bed.

The funniest thing about him is the way he likes to grow –
Not at all like proper children, which is always very slow;
For he sometimes shoots up taller, like an india-rubber ball,
And he sometimes gets so little that there's none of him at all.

He hasn't got a notion of how children ought to play,
And can only make a fool of me in every sort of way.
He stays so close beside me, he's a coward you can see;
I'd think shame to stick to nursie as that shadow sticks to me!

One morning, very early, before the Sun was up,
I rose and found the shining dew on every buttercup;
But my lazy little shadow, like an arrant sleepy-head,
Had stayed at home behind me and was fast asleep in bed.

Robert Louis Stevenson

HOW LONG IS A

A Year 3 class placed a metre stick in a lump of clay on their school playground. Every hour they measured the length of the shadow. They made their results into a graph.

SHADOW?

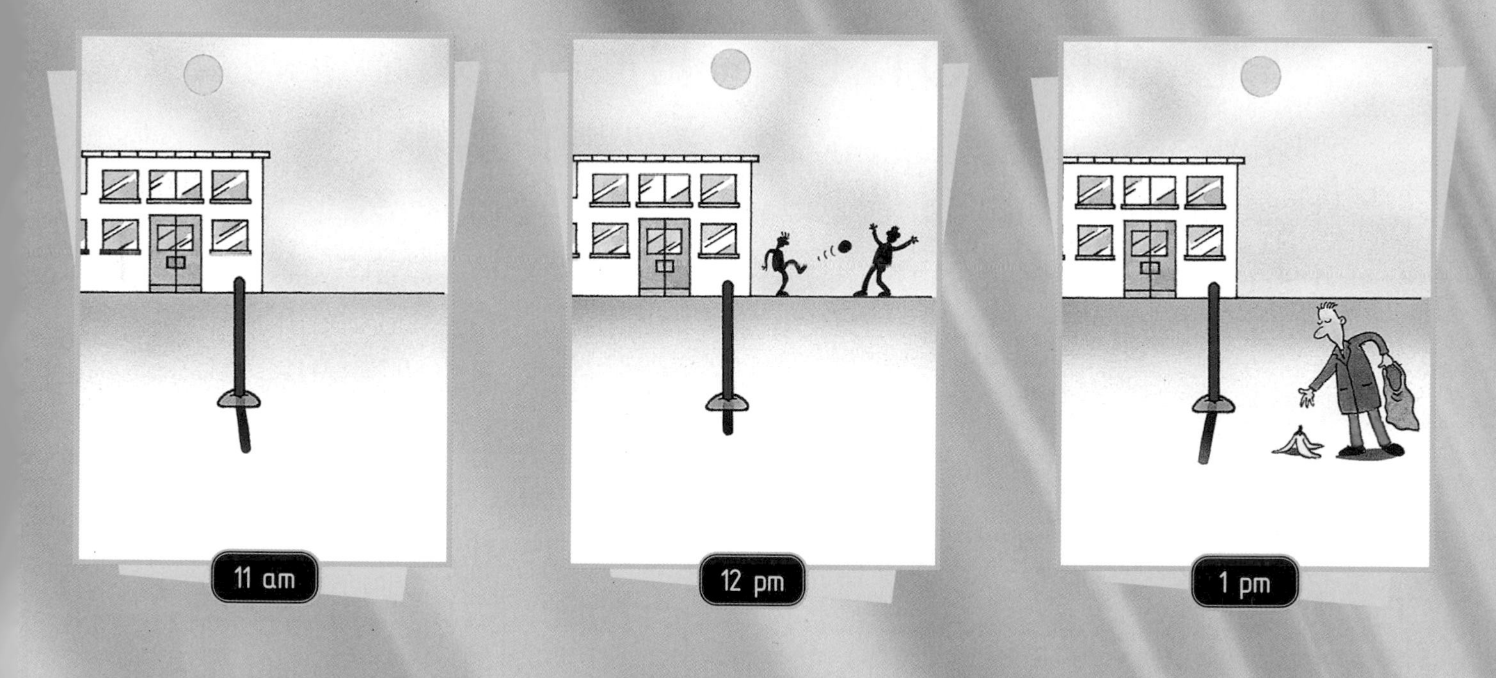
11 am
12 pm
1 pm

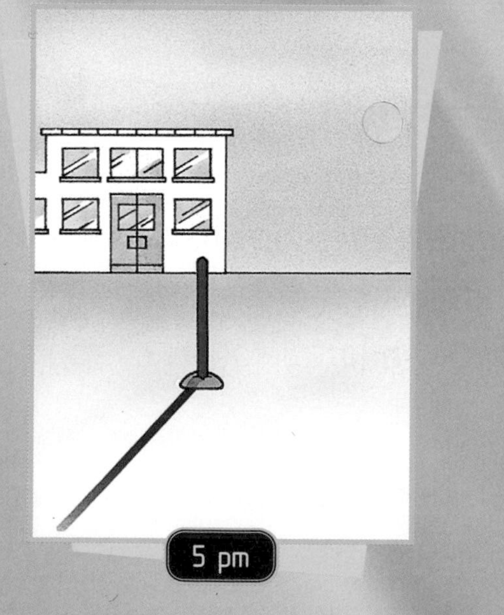
5 pm

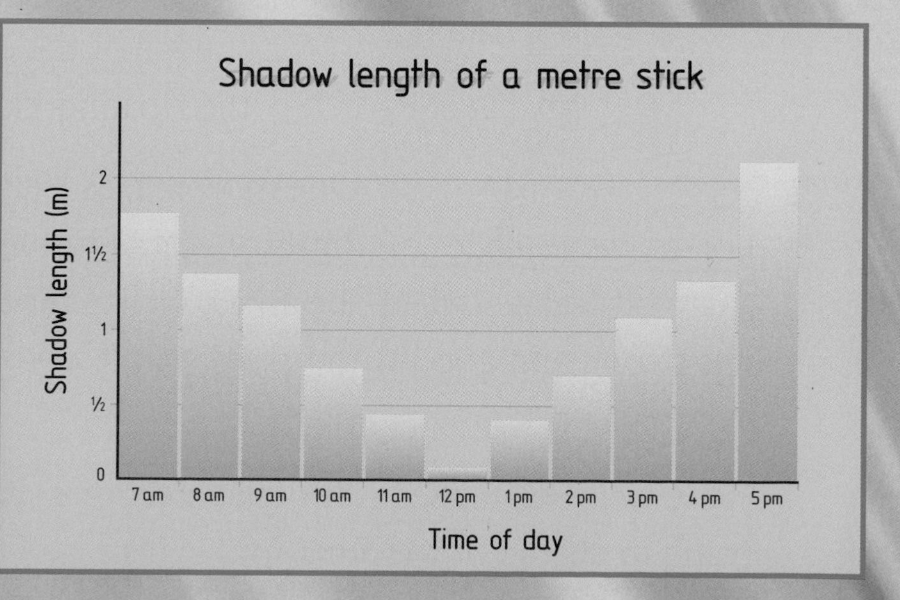
Shadow length of a metre stick
Shadow length (m)
0
½
1
1½
2
7 am
8 am
9 am
10 am
11 am
12 pm
1 pm
2 pm
3 pm
4 pm
5 pm
Time of day

FROM EAST TO WEST

Thousands of years ago, before the first scientists began to question how and why the Sun seemed to travel across the sky, people told each other stories to explain how it rose every day in the east and set in the west.

THE CHARIOT OF THE SUN

The Ancient Greek god of the Sun was called Helios. He lived on a floating island which drifted backwards and forwards across a great ocean. Every morning, the island started at the very eastern edge of the ocean. Helios would rise from his bed and harness his mighty horses to a chariot that held the Sun. Then he would drive the chariot westward across the sky, bringing light and warmth to the world. At the end of the day, Helios guided the horses down to the western edge of the great ocean to where the floating island had drifted.

One morning, as Helios harnessed the horses, his teenage son, Phaeton, came rushing over. "Why not have a day off today, father?" he asked. Helios thought about it. Phaeton was strong and good with horses, and it would be wonderful to have a day off work!

"Very well," he replied. "But take great care. It's an important job, being in charge of the Sun."

Phaeton climbed into the chariot, and cracked his whip to set the horses galloping across the sky. To begin with he thought it was exciting to race across the heavens like this. However, as the day wore on, Phaeton grew bored.

"I wonder what would happen if wc went a little higher," he muttered to himself. With a crack of the whip, Phaeton sent the horses swooping upward. Now it was exciting again! Up and up they went, higher and higher. But soon the Sun was so high in the sky that its heat and light no longer reached the Earth. Darkness fell across the land and sea, like a terrible wintry midnight. Everything turned to ice and people huddled together in terror. It seemed as if all mankind would die.

Luckily, Zeus, the king of the gods, was watching. He hurled a thunderbolt to knock Phaeton from the chariot and send him plunging down to Earth. The horses stopped their wild climb and swooped back down to their usual path across the sky. At last they found their way home and landed on the floating island at the western edge of the Earth.

And from that day to this, Helios has never again let anyone but himself drive the Chariot of the Sun.

THE RABBIT AND THE SUN
Ha! Hot is good!
It's too hot! Get back into the sky, Sun

Sun has never been the same. Every morning he peeps carefully over the horizon, just to see where Rabbit might be. Then, when he is in the sky, he makes sure he is always too high to be hit by another arrow.

THE EARTH MOVES

Ptolemy

(2nd Century AD)

- Greek scientist – did not believe myths
- In 140 wrote book (the Almagest)
- Said Earth was centre of Universe – Sun moved round the Earth
- The Christian Church agreed with him
- His ideas accepted for almost 1500 years

Galileo

(1564–1642)

- Italian scientist
- 1609 – made **telescope** – observed Venus
- Proved Venus circled Sun –other planets did so too
- 1635 Church leaders refused to accept Galileo's ideas and made him say he was wrong

Copernicus

(1473–1543)

- Polish **astronomer**
- Observed planets – worked out they were going round Sun
- Thought Earth must go round Sun too
- 1543 wrote book (De Revolutionibus)
- Christian Church banned book because it went against the Bible

Johannes Kepler

(1571–1630)

- ★ German astronomer – observed night sky
- ★ Wanted to prove Copernicus right
- ★ 1609 and 1619 wrote books based on observation of planets and how they moved
- ★ Supported Galileo's ideas

Issac Newton

(1642–1727)

- ★ British scientist
- ★ Accepted Galileo's ideas (the ideas were accepted by everyone)
- ★ 1687 – proved that gravity kept planets circling round Sun
- ★ 1704 wrote book about light
- ★ Became very famous – 1705 became Sir Isaac Newton

THE LIFE OF GALILEO

Galileo was a scientist who was born in Italy in 1564. He believed that careful observation and experiments would help him find out about the Universe and how it worked. However, the powerful men in charge of the Christian Church at that time did not like what he found.

In 1609 Galileo made one of the first **telescopes** and used it to look deep into space. He discovered the moons of Jupiter and the rings of Saturn. He also used his telescope to observe the planet Venus more closely than anyone before him.

Galileo noticed that Venus seemed to change its shape, just as the Moon did. He realized that these different shapes must be caused by the Sun shining on Venus, as Venus moved around it. If Venus was moving around the Sun, all the other planets – including the Earth – must be doing the same. Galileo's observations showed that the Sun was the centre of the **solar system**.

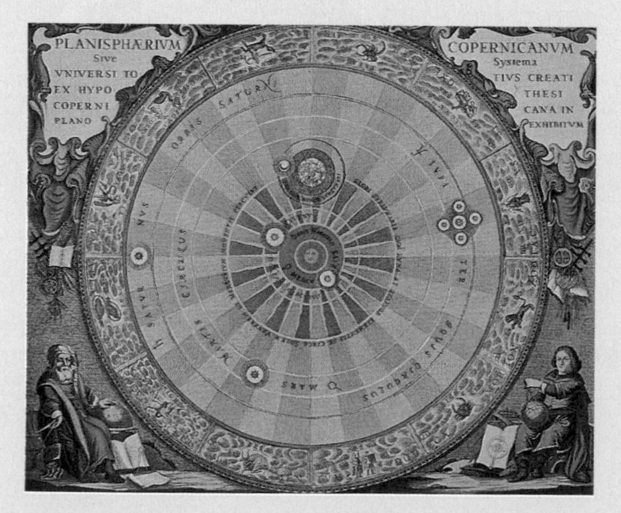

An early map of the solar system

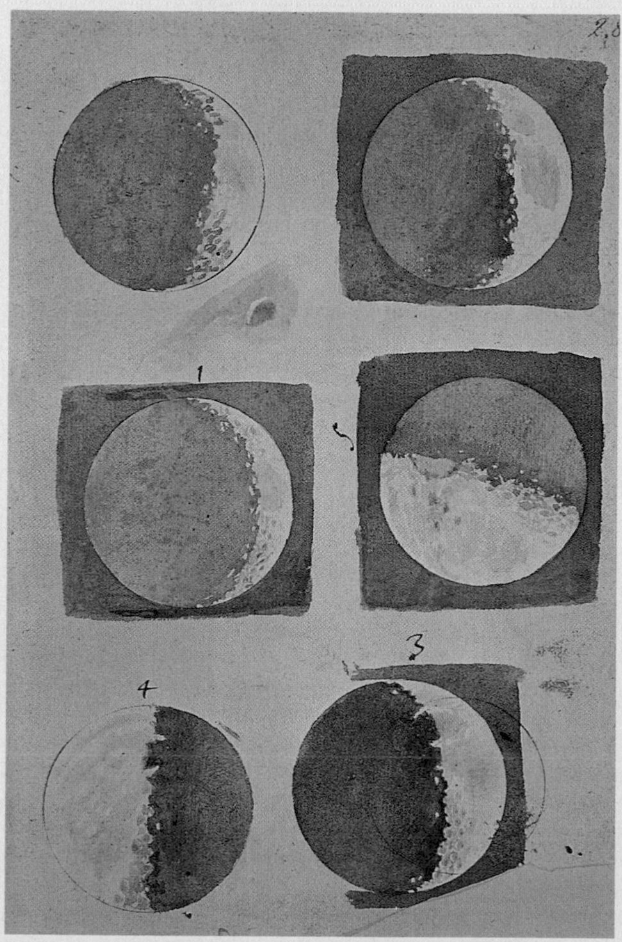

Galileo's drawing of Venus

However, the leaders of the Church still believed that the Earth stood still and the Sun, Moon and stars moved around it. They did not like what Galileo was saying, and they refused to accept his scientific evidence. In 1635, the Church leaders told Galileo that he had to make a simple choice. Either he could say he was wrong and save his life, or he could go on saying he was right and be tortured and executed!

'Trial of Galileo' by Nicolo Barabina (1832–1891)

Galileo said he was wrong. The statement saved his life, but the Church leaders still arrested him and kept him as a prisoner in his own house. He lived for another seven years and died in 1642 when he was 77 years old.

THE SUN STAYS STILL

What you need

- A table lamp without a lampshade
- Some Plasticine
- A globe that rotates

1

Make a tall column of plasticene and stick it on the place where you live.

2

Turn the globe so the place where you live is on the side furthest from the Sun. Switch on the light. The Earth spins once every twenty-four hours. When the place where you live is on the opposite side of the Earth to the Sun, it is night. There is no Sun and no shadow.

3

Turn the globe so that it spins slowly anti-clockwise so countries like Japan catch the light first. (North is at the top of the globe and South is at the bottom. Anti-clockwise around the globe is East and clockwise is West.)

Watch what happens as the place where you live gets closer to the Sun. The Sun seems to rise in the East. The shadow cast by the column starts off long and stretches to the West.

When the Sun is overhead there is little or no shadow.

As the Earth turns away from the Sun, it seems the Sun is going down in the West. The shadows grow longer and stretch out to the East.

The Sun stays still, but the Earth keeps turning. When it has turned away from the Sun, it is night time once again.

ECLIPSE!

The Sun's diameter is about 400 times the diameter of the Moon. However, the distance from the Earth to the Sun is about 390 times the distance from the Earth to the Moon so the Sun looks about the same size as the Moon in the sky. This means when the Moon passes between the Earth and the Sun it blocks its light and makes a shadow fall on the Earth. This is called a **solar eclipse**.

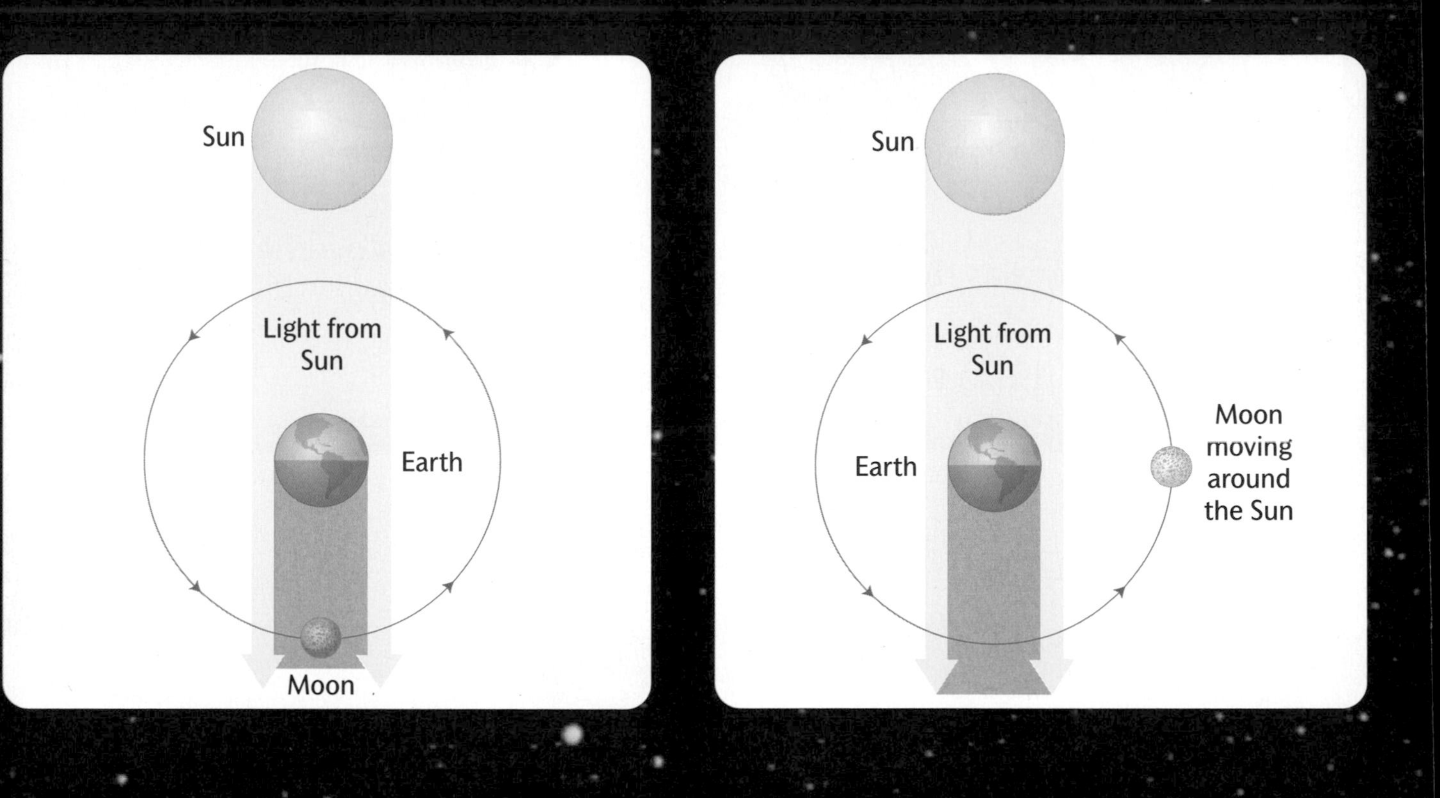

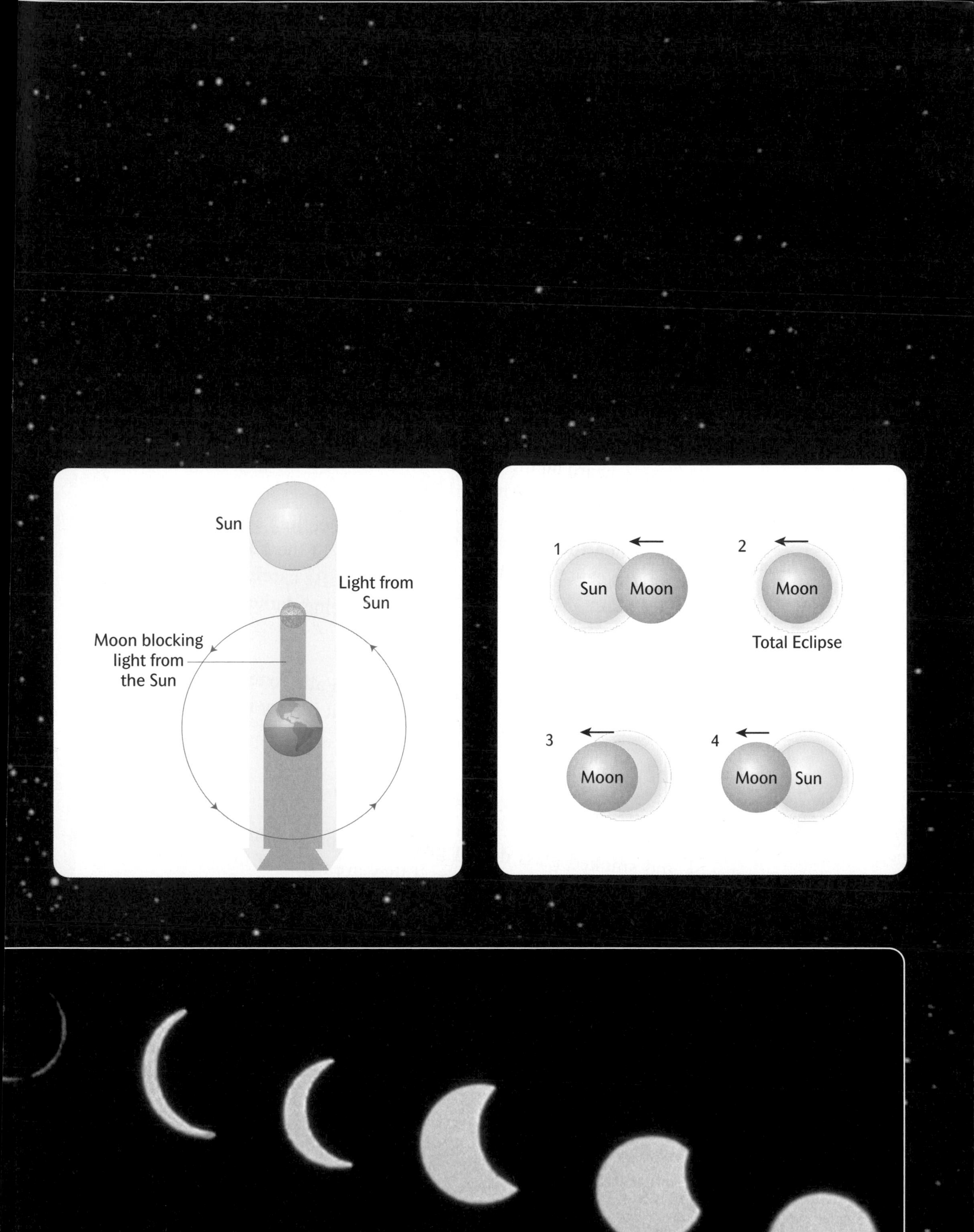
Sun
Light from Sun
Moon blocking light from the Sun
1
Sun
Moon
2
Moon
Total Eclipse
3
Moon
4
Moon
Sun

THE SCHOOLS LIAISON SERVICE
SCIENCE DEPARTMENT NORTHWESTBURY AY8 6SJ

To all Pupils,
Northwestbury Schools

July 10th 2000

Dear Pupils,

As you may well have heard on the News, or from your teachers, a total eclipse of the Sun is forecast for August 11th.

An eclipse occurs when the Moon blocks the light travelling from the Sun to the Earth. This causes a shadow on the surface of the Earth. For a short time it will seem as if evening has come early. Be careful if you are out cycling, or playing near a road. As well as being darker it may well be colder too.

Although this is, of course, an exciting event for you all – one that you are unlikely to see repeated again in your lifetime – it is very important to follow the 'Eclipse Code':

NEVER look directly at the Sun.

Sunglasses and smoked glass do not give proper protection. Welders' masks or goggles can protect your eyes if they are made of the correct materials, but the safest method to use is the pinhole camera, as described in the Teacher's Pack that accompanies this letter.

Safe viewing!

Yours sincerely,

Roberta Blake

Ms Roberta Blake
KS2 Science adviser

MORGAN, Jo

From: TAYLOR, Jake
Sent: Saturday, 12th August, 2000 11:30 am
To: MORGAN, Jo
Subject: Eclipse!

Jo,

Wow! Did you see it? Amazing!.... We didn't get it complete, though. Before the Sun was completely covered some clouds came over, so Mum went inside and watched it on TV. I stayed out and then the clouds cleared enough to see it at the end. I was lucky because it meant I used Mum's welder's goggles so I could look at the Sun without hurting my eyes.

When it got really dark the birds stopped twittering and people over the road started cheering. The light was really weird, a dim, greyish colour.

Anyway, I'm glad I saw it – it was a once in a lifetime thing.

Jake

THE MOON AT NOON

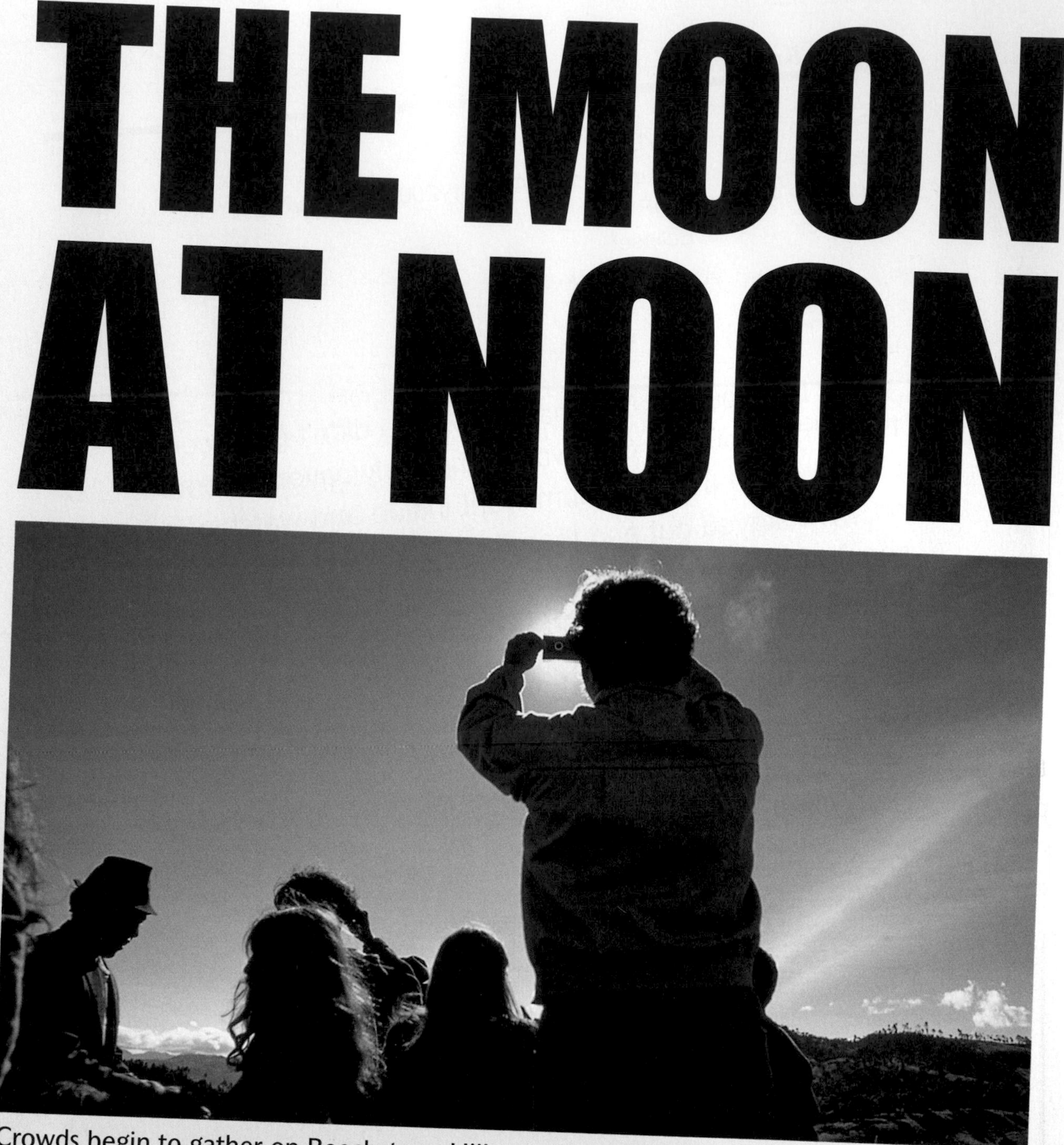

Crowds begin to gather on Beechstone Hill

By SIÂN MAY
Science Correspondent

A SIGHT not seen for 70 years brought towns and villages across Britain to a halt this afternoon.

Instead of cutting the grass, walking the dog, reading the paper, playing cricket, setting out for a drive or enjoying any other of their usual sunny Summer afternoon activities, the people of Beechstone in the South West came out into the streets and stared at the sky.

Danger

Fortunately most of them had listened to the warnings and followed the Eclipse Code. Nobody tried to look at the Sun through sunglasses, though, Mr Graham, the landlord of the village pub, "The Lazy Sheepdog", reported a steady sale of smoked glass. He said, "if it was safe for my Grandad to use it in 1936, it's still safe to use it now."

Total eclipse!

Thrilling

When the Moon finally slipped across the face of the Sun at 11.13a.m. precisely, a gasp went up from the waiting crowd. "Thrilling, just amazing!" said Jennie Ebrahim, aged 10.

Corona

What was unusual about this eclipse was the fact that it was total. The circle of the Moon covered the face of the Sun completely. Apart from the corona – the flaring edge of the Sun – the face of the Sun was completely blocked out.

Spooky

Replacing the bright light of noon with the grey light of dusk affected the human inhabitants of Beechstone more than the animals. The birds carried on singing, the sheep and cows did not lie down and go to sleep, and the dogs did not bark. But for the few minutes the total eclipse lasted the people of Beechstone were strangely quiet. "It was quite spooky," remarked Dylan Protheroe, the vicar of St. Jude's, "almost as though the end of the world had come."

Night or Day?

PRISONERS OF THE SUN

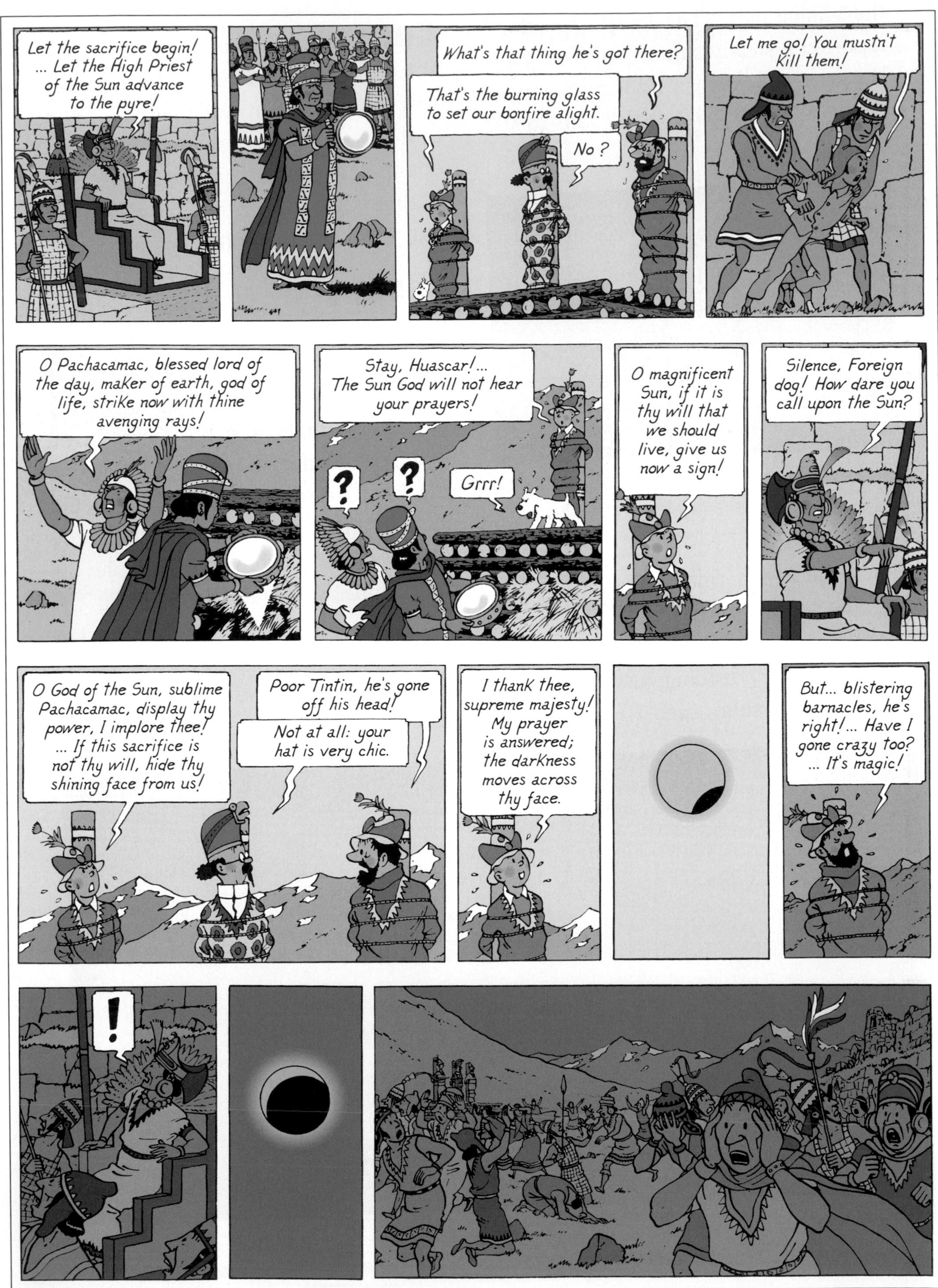

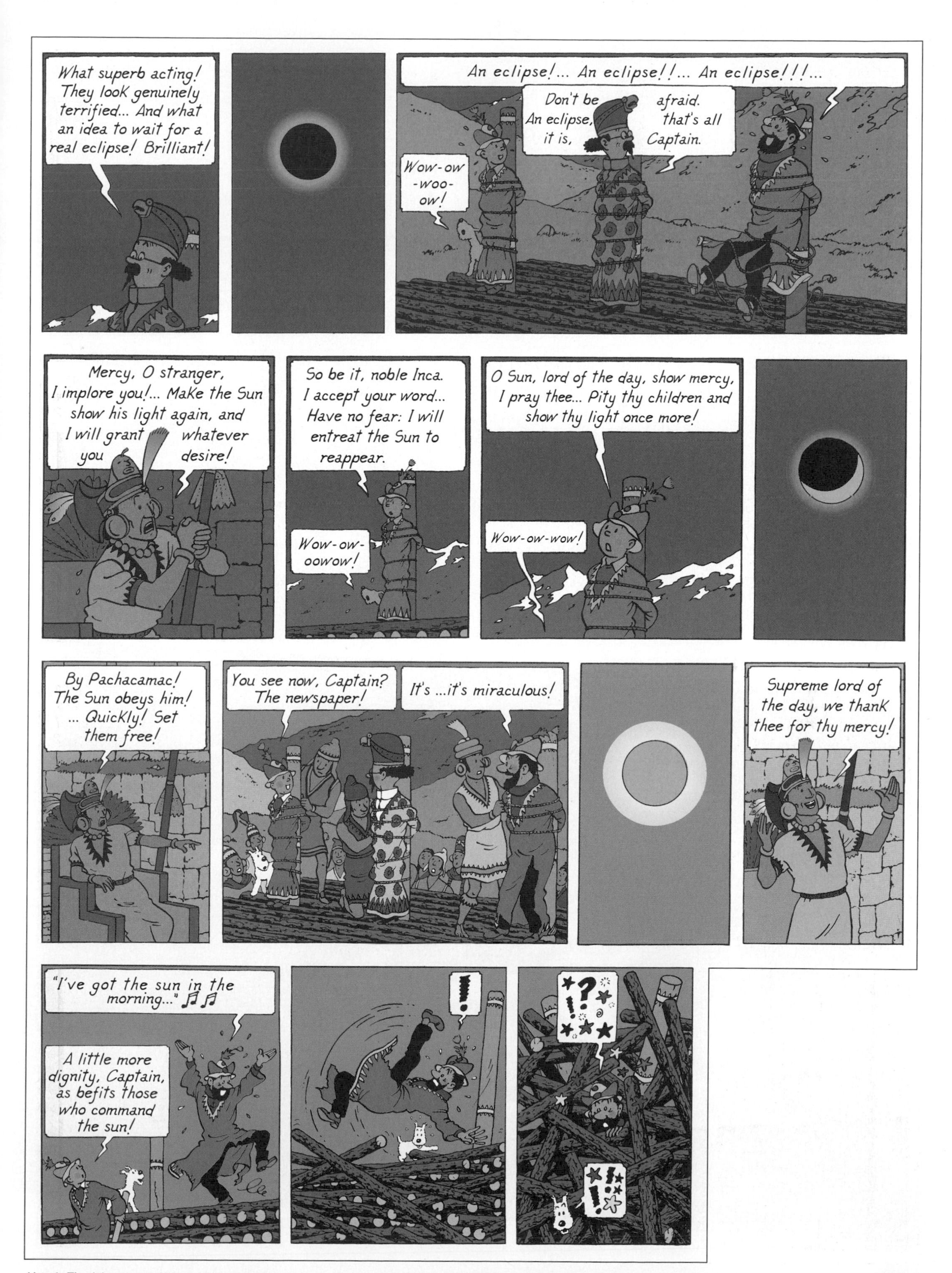
What superb acting! They look genuinely terrified... And what an idea to wait for a real eclipse! Brilliant!
An eclipse!... An eclipse!!... An eclipse!!!...
Don't be afraid. An eclipse, that's all it is, Captain.
Wow-ow -woo- ow!
Mercy, O stranger, I implore you!... Make the Sun show his light again, and I will grant you whatever desire!
So be it, noble Inca. I accept your word... Have no fear: I will entreat the Sun to reappear.
Wow-ow- oowow!
O Sun, lord of the day, show mercy, I pray thee... Pity thy children and show thy light once more!
Wow-ow-wow!
By Pachacamac! The Sun obeys him! ... Quickly! Set them free!
You see now, Captain? The newspaper!
It's ...it's miraculous!
Supreme lord of the day, we thank thee for thy mercy!
"I've got the sun in the morning..."
A little more dignity, Captain, as befits those who command the sun!
!
?!
!!

LIGHT IN THE DARK

Throughout history, light has been an important religious symbol. Because light is so important in their lives, people have always connected it with the most powerful forces in their world.

STONEHENGE

For the people who built Stonehenge over 3500 years ago, the rising Sun was a symbol that life would return to the Earth after the cold and darkness of winter.

They had such a good knowledge of the Sun's movements across the sky they were able to put the giant stones in exactly the right places so that they could act as markers.

The sun rises over Stonehenge on Midsummer's Day

A nine-branched candlestick used during Hanukkah

JUDAISM

Hanukkah – a festival of light – helps Jews to remember the story of a miracle that happened in Jerusalem many years ago.

The Jews took back the city after invaders had tried to destroy the Old Temple. When the Temple was being made ready for prayer they noticed that there was only enough oil in the lamp for one day. In order for them to pray, the lamp was to be lit at all times. They were upset because they knew it would take a week to get more oil. However, that small amount lasted eight days. A miracle!

Today, during Hanukkah, a candle is lit for each day the oil lasted. One extra candle is used to light all the others. This is why there are nine spaces on the special candlestick or Hanukiyah. The festival of light is a reminder of faith in difficult times.

Hinduism

The Hindu festival of **Divali**, known also as the Festival of Lights, celebrates the Hindu New Year. The very first Divali was held to celebrate the return home of Prince Rama and his wife Sita. After being banished from their home for fourteen years, Sita was kidnapped by the demon king Ravana. But Rama, with the help of the monkey King Hanuman, rescued her. When they finally returned home it was night time, so the people lit lamps to guide their way. Today, Hindus still celebrate by putting lights in their windows to remind them of Rama defeating Ravana. They also believe that Lakshmi, the goddess of good luck and wealth, will bring fortune to the homes that are well lit with lights and candles.

Light has been used as a symbol in many other religions.

The Ancient Egyptians had three gods – Horus, Ra and Osiris – who represented the rising, noontide and setting Sun.

Christians call Jesus 'The Light of the World'.

The Vikings said the Sun was the all-seeing eye of Odin.

SHADOW TIME

A **sundial** is a timekeeper that is powered by the Sun. It has a **gnomon** which sticks up and casts a shadow on the hour marks. The shadow is like the hand on a clock: as it moves it points to the correct time.

On most garden sundials the gnomon points to the north. As the Sun rises in the sky, the shadow creeps round the hour scale. At noon, the Sun is directly overhead and there is little or no shadow.

A sundial cannot work properly unless it is pointing in exactly the right direction. This means that most sundials have to be fixed in one place. However, portable sundials were first invented by the Romans almost 2000 years ago.

Egyptian Shadow Bar

This is one of the earliest examples of a sundial. The one in the picture was made in about 900 BC. It is about 30 cm long.

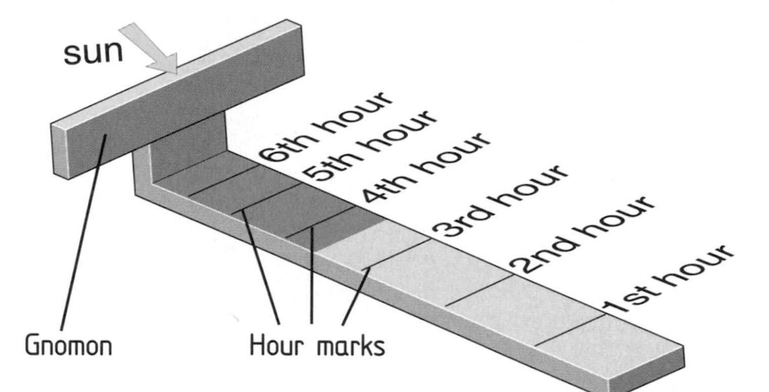

At sunrise, the T-shaped end is pointed towards the East. As the Sun rises in the sky the shadow creeps along the hour scale towards the T. At noon, the Sun is directly overhead and there is no shadow. During the afternoon, the shadow bar is turned around. The shadow now tells the hours by creeping away from the T.

By 1600 AD clocks and watches told the time much more accurately than sundials, and they worked whatever the weather – and all through the night! But sundials have never gone out of fashion.

Diptych dials

- Folded up – so portable
- Very popular in 16th century
- Made of ivory – expensive
- Often have compass inside to find the right direction

Analemmatic dials

- Gnomon is a human!
- Must face north and stand still
- Must stand very straight

Sundials come in all shapes and sizes

Garden sundials

- Used as garden ornaments
- Popular in Victorian times
- Often had mottoes carved on them

Hour mark

Gnomon

Let others tell of storms and showers, I'll only count your sunny hours.

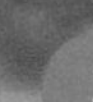

HOW TO MAKE A

SHADOW PUPPET

4

5

6

WHAT LETS LIGHT

What lets light through?

THROUGH?

Blackout!

During the Second World War (1939–1945) aircraft often attacked large cities and towns. To make it more difficult for the bombers to find their targets, at night time people covered their windows with **opaque** blinds and curtains. They were called blackout curtains. This meant that the lights inside the house couldn't be seen from the air.

DIFFERENT MATERIALS, DIFFERENT SHADOWS

Quiz

Can you guess what objects made these shadows?

How many words can you think of to describe each shadow?

Why are the shadows different?

What do you think was the light source each time?

Can you tell whether the light source was high up or low down?

GLOSSARY

astronomer a person who studies the Sun, moons, planets and stars

Divali a Hindu festival of lights held in November

eclipse when the moon moves between the Sun and the Earth

gnomon the pointer on a sundial that casts the shadow which tells the time

Hanukiyah a nine - branched candlestick

Hanukkah a Jewish festival of lights held in November or December

opaque something that does not let light through

silhouette a dark shadow seen against a light background; the shape and outline of an object usually in black

solar system the planets and their moons which move around the Sun

sources where something comes from or begins

sundial a way of telling the time using shadows

telescope an instrument to make distant objects appear bigger

translucent lets some light through

transparent lets a lot of light through

BIBLIOGRAPHY

Non-fiction

Allday *The Young Oxford Library of Science Light and Sound*
ISBN 0-19-910946-X

Bevan *Sacred Skies: The Fact and the Fable (Landscapes of Legend),* ISBN 0749625511

Jennings, T *Science Success* Book 1
ISBN 0-19-918338-4

Jennings, T *The Oxford Children's A-Z to Science*
ISBN 0-19-910992-3

Langley, A *Oxford First Encyclopaedia*
ISBN 0-19-910974-5

Morgan *Oxford First Book of Science*
ISBN 0-19-910501-4

Taylor and Pople *The Oxford Children's Book of Science*
ISBN 0-19-910084-5

Shadow Games: A Book of Hand and Puppet Shadows
ISBN 1570540306

The Young Oxford Encyclopaedia of Science
ISBN 0-19-910711-4

The Oxford Children's Illustrated Encyclopaedia
ISBN 0-19-910444-1

Fiction

Greek and Norse Legends The Illustrated Guides
ISBN 0746002408

Stevenson, R.L. *My Shadow*
ISBN 0879237880

Internet

www.scienceyear.com/home.html

www.scienceweb.org.uk/html/f_3f.htm

www.learn.co.uk/default.asp?WCI=Unit&WCU=11301

www.learn.co.uk/default.asp?WCI=Unit&WCU=11288

www.uffington.oxon.sch.uk/class1/lightdark.html

http://www.cheltenhamfestivals.co.uk/frame_festindex.cfm?FEST=SCIENCE

Organizations

At-Bristol, Harbourside, Bristol
http://www.at-bristol.org.uk/

Science Museum, London
www.sciencemuseum.co.uk

INDEX